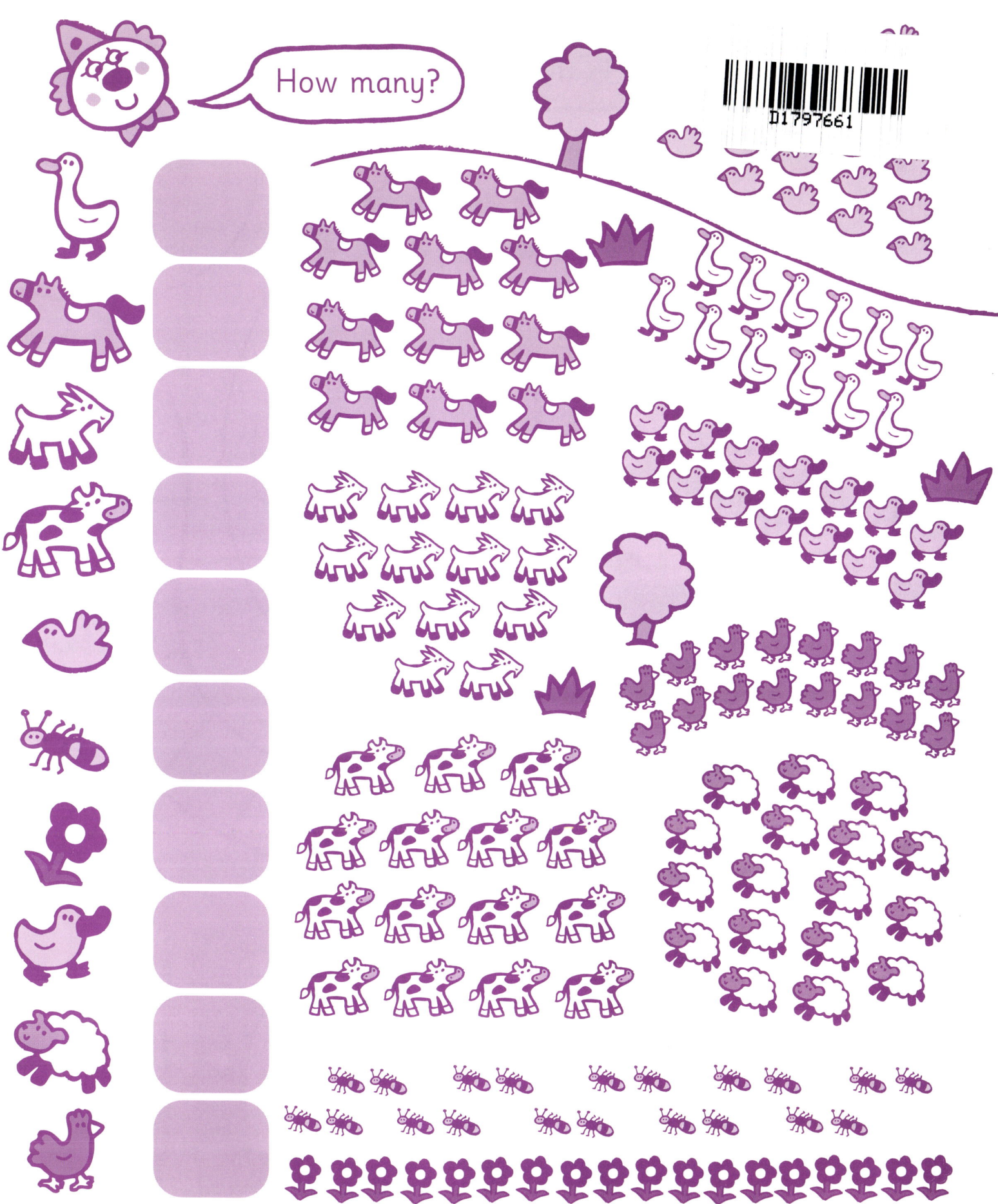

How many?

Put in the numbers.

I am ___ years old

Choose numbers.

Before you start
IP 8

• Can write numbers to 20 / ___ .
Notes/date:

First Skills in Numeracy 1

# Practice Book 3

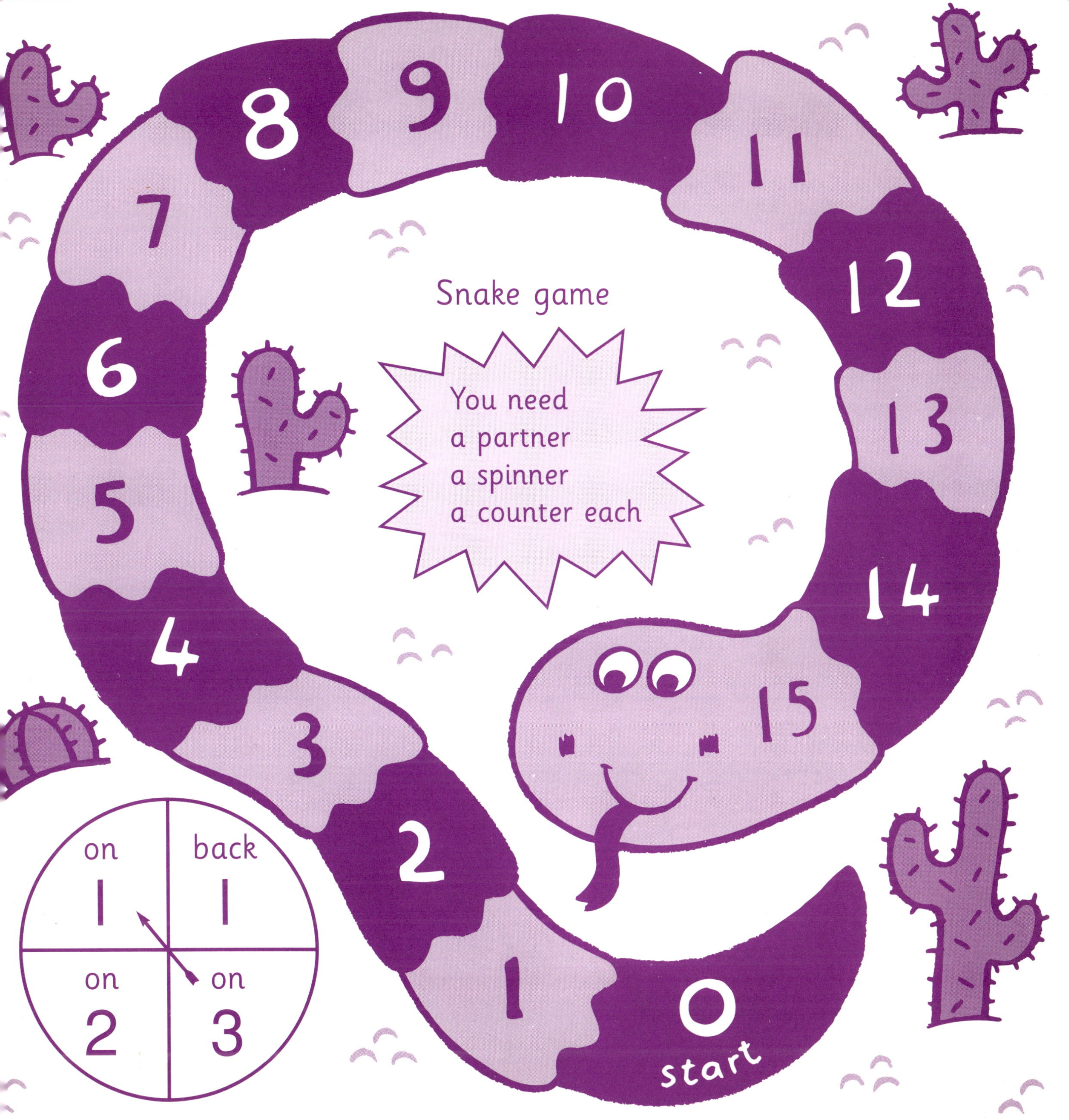

Snake game

You need
a partner
a spinner
a counter each

8 9 10 11 12 7 6 13 5 14 4 3 15 2 1 0 start

| on | back |
| 1 | 1 |
| on | on |
| 2 | 3 |

Do the beads game with your teacher.

How many? ☐

How many? ☐

How many? ☐

How many? ☐

Draw ☐

Draw ☐

Put in the numbers.

I add 3 is ☐

Do more.

Before you start
IP 14

● Can visualise numbers in 5s and a bit more.
● Can add on the number line.
Notes/date:

Splitting numbers

Can you see the number pattern?

2 and 3 is 5

2 + 3 = 5

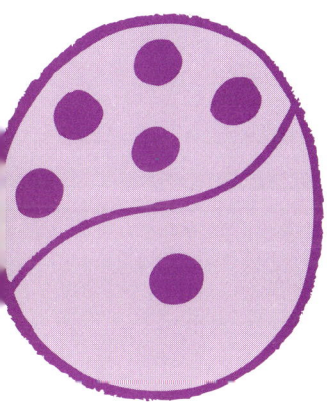

5 and 1 is 6

5 + 1 = 6

Choose numbers.

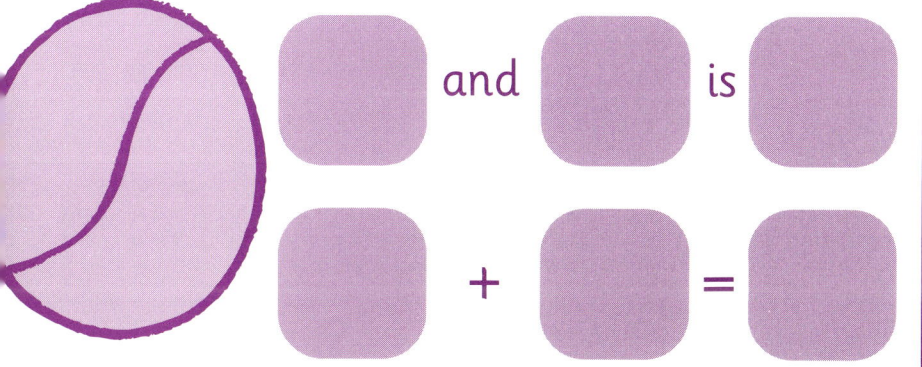

and [ ] is [ ]

[ ] + [ ] = [ ]

2 + 0 = [ ]

2 + 1 = [ ]

2 + 2 = [ ]

2 + 3 = [ ]

2 + 4 = [ ]

2 + 5 = [ ]

2 + [ ] = [ ]

2 + [ ] = [ ]

● Can split numbers to 10 / above.
● Can add pairs of numbers to 10 / above.
Notes/date:

Before you start
IP 5, 13
CG Splitting numbers

5

You need: cards 0–10                                                cubes

Make sums

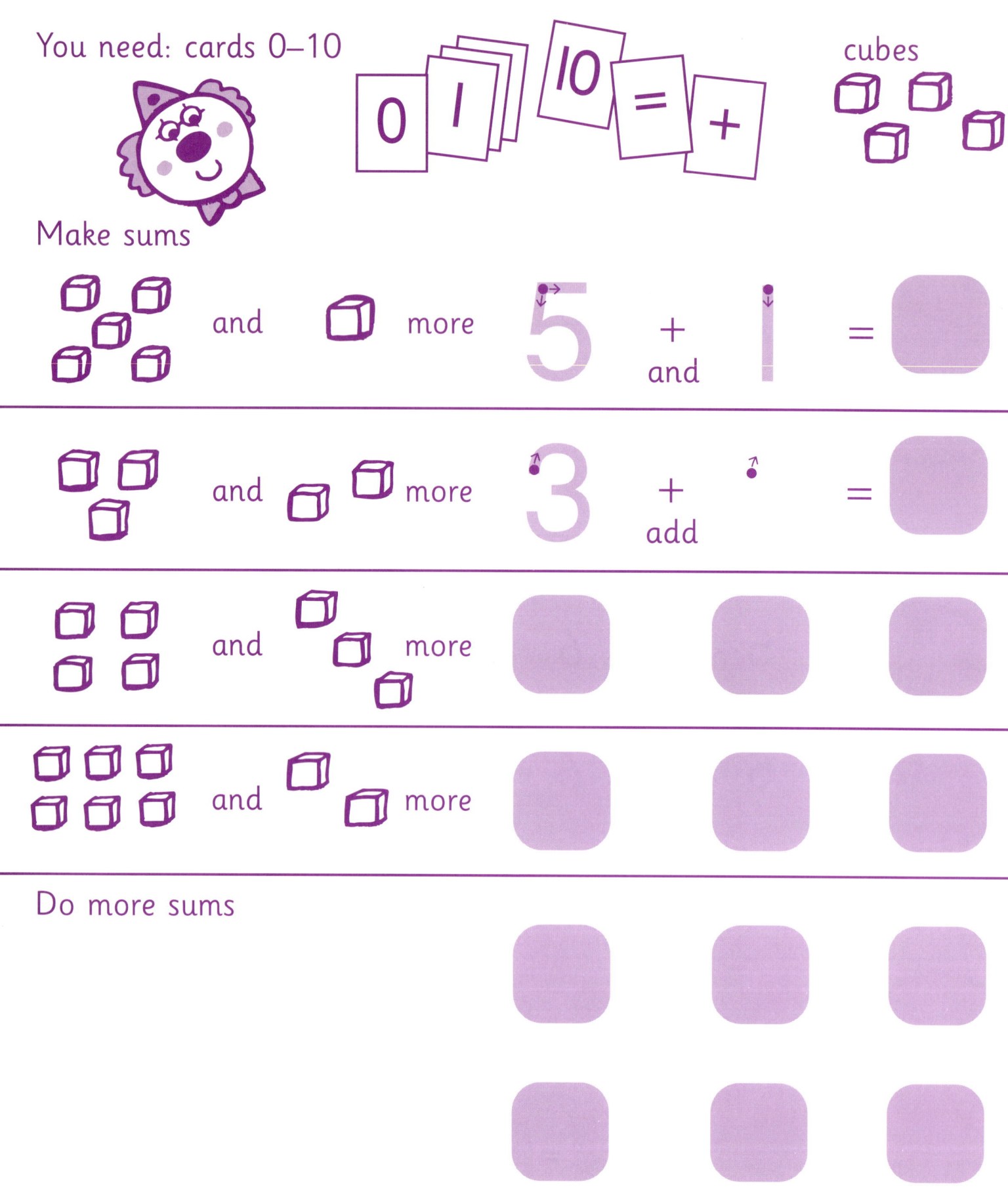

and          more          5     +     1     =
                                       and

and      more          3     +           =
                            add

and          more

and          more

Do more sums

     ● Can add with cubes to __.
     Notes/date:

You need: cards 0–10   cubes

$3 + 2 =$ ☐

$1 + \phantom{0} =$ ☐

$2 + \phantom{0} =$ ☐

$\phantom{0} + \phantom{0} =$ ☐

Choose numbers

 In my head

- Can add to ___ with fingers/cubes.
- Can add mentally to ___ .
Notes/date:

Before you start
IP 13

7

In your head

In my head
I can

add **1** to a number

1 add **1** → 2

2 add **1** → 3

3 add **1** →

4 add **1** →

5 add **1** →

add **2** to a number

1 add **2** →

2 add **2** →

3 add **2** →

add **2** →

add **2** →

add ☐ to a number

1 add ☐ →

2 add ☐ →

3 add ☐ →

Try more

Before you start
IP 15, 13
CG In your head

● Can add 1, 2, __ mentally to numbers up to __ .
Notes/date:

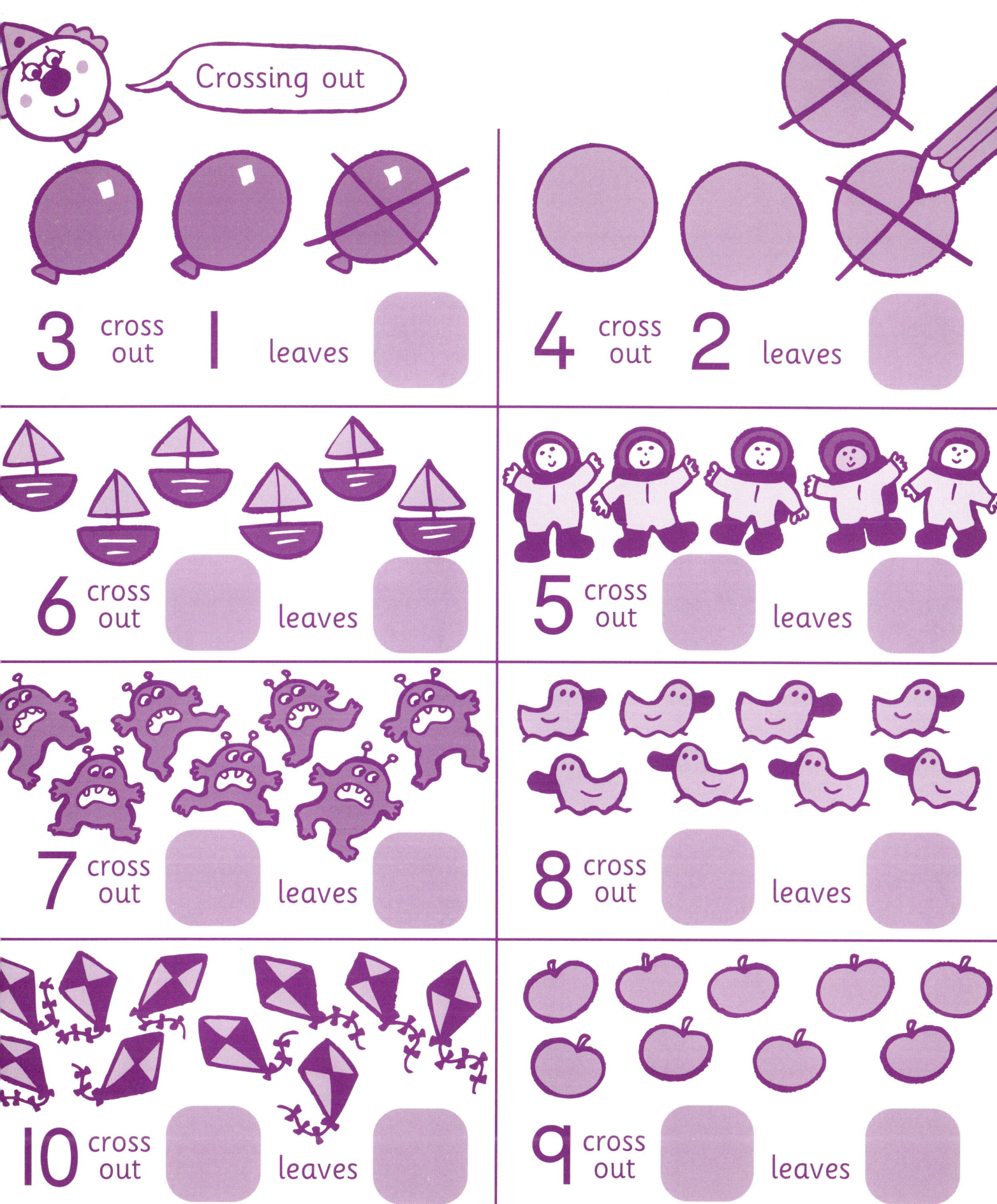

Crossing out

3 cross out 1 leaves ☐

4 cross out 2 leaves ☐

6 cross out ☐ leaves ☐

5 cross out ☐ leaves ☐

7 cross out ☐ leaves ☐

8 cross out ☐ leaves ☐

10 cross out ☐ leaves ☐

q cross out ☐ leaves ☐

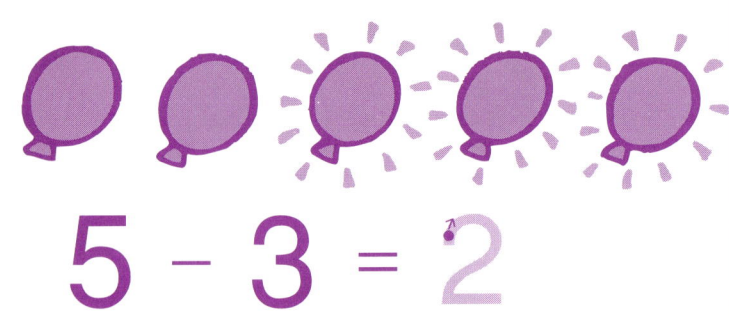

$5 - 3 = 2$

$4 - 3 = $

$5 - 2 = $

$7 - 2 = $

$10 - 2 = $

$9 - \phantom{0} = $

Cross out

4 cross out 1 leaves

Draw

cross out leaves

Draw more

Before you start
IP 11

● Can subtract from numbers to __ .
Notes/date:

Take away

You need: cubes  dice

cards 0–10

3 take away 2 =

8 take away 5 =

7 take away [ ] =

10 take away [ ] =

In my head

● Can subtract from numbers to 10 with fingers/cubes.
● Can subtract mentally from numbers to __ .
Notes/date:

Before you start
IP 11
CG Take away

11

Start with 10 mice.
Draw what you did.

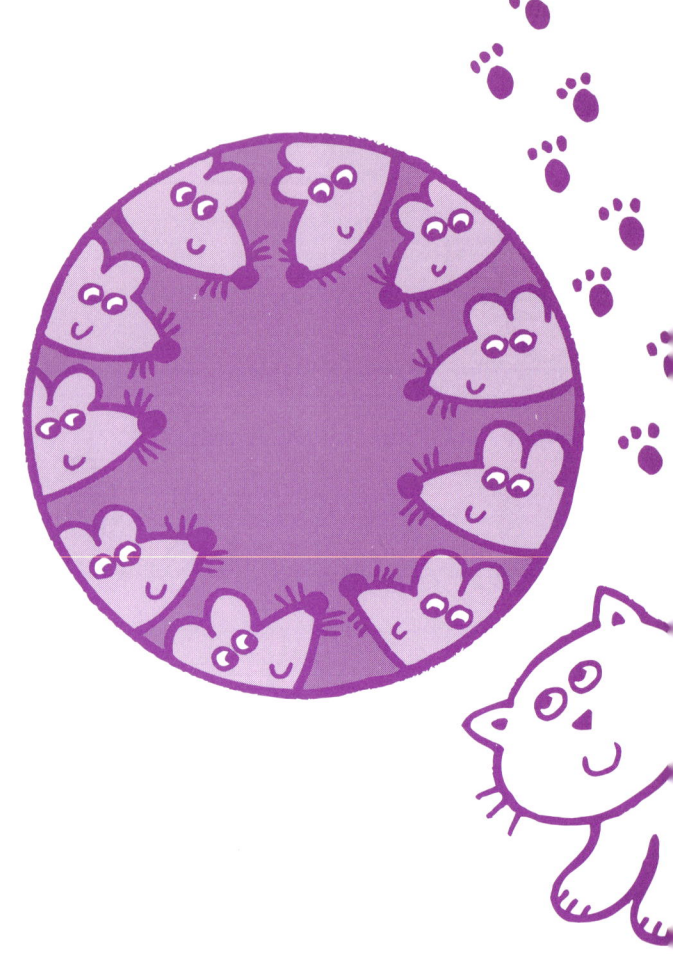

Before you start
IP 13

● Can record number bonds to 10.
Notes/date:

Make up to 10

Make all the dominoes up to 10.

Join pairs to make 10.

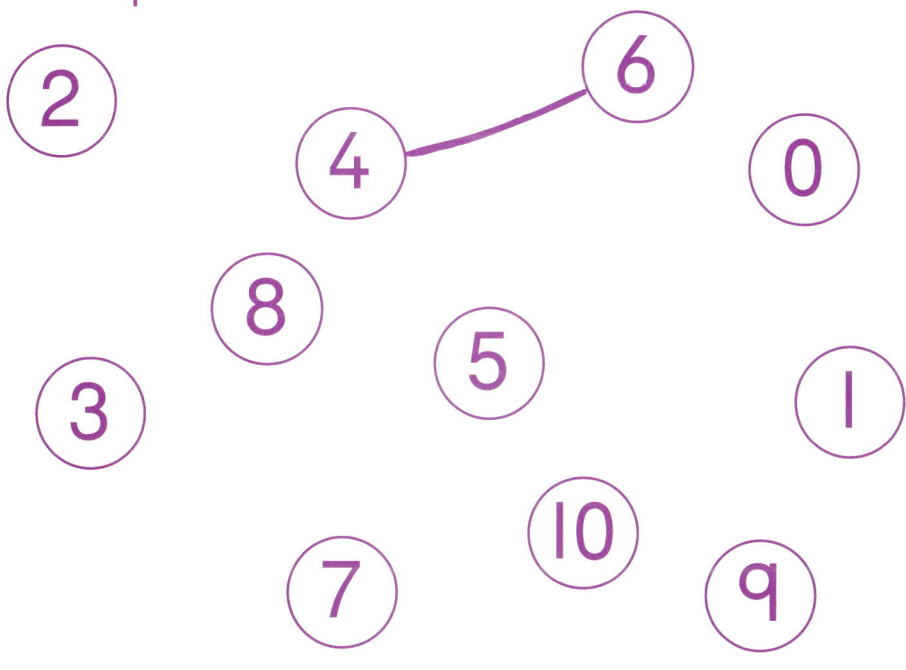

0 + 10 =

1 + 9 =

2 + 8 =

3 + 7 =

4 + = 10

5 + = 10

6 + = 10

7 + = 10

8 + = 10

9 + = 10

10 + = 10

Can you see the number pattern?

• Knows number bonds to 10.
Notes/date:

Before you start
IP 7, 13
CG Make up to 10

13

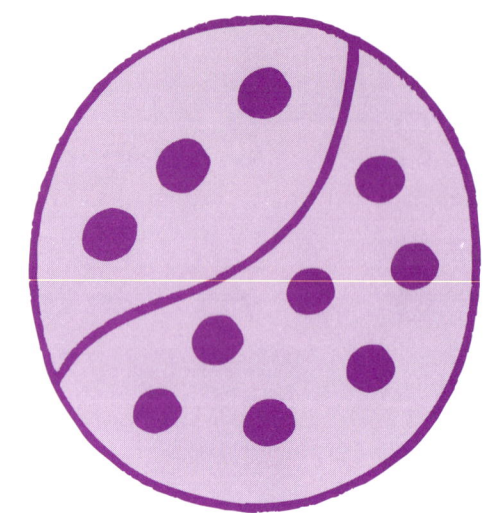

3 and 7 is 10

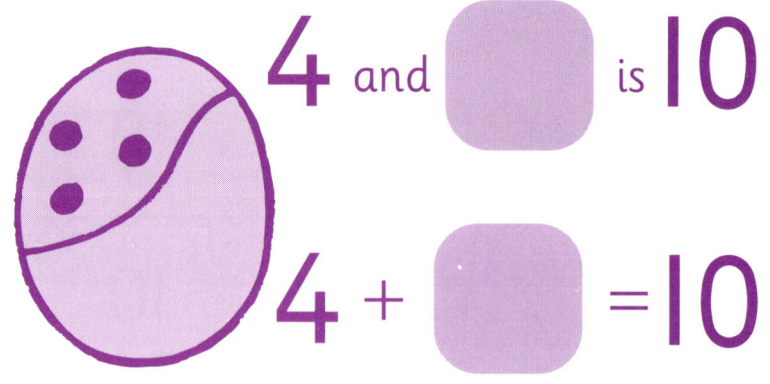

4 and ☐ is 10

4 + ☐ = 10

2 and ☐ is 10

2 + ☐ = 10

Choose numbers

☐ and ☐ is ☐

☐ + ☐ = ☐

☐ and ☐ is ☐

☐ + ☐ = ☐

Before you start
IP 13, 16
CG Splitting numbers

● Can split numbers to 10 / above.
Notes/date:

PUBLISHED BY THE PRESS SYNDICATE OF THE UNIVERSITY OF CAMBRIDGE
The Pitt Building, Trumpington Street, Cambridge, CB2 1RP, United Kingdom

CAMBRIDGE UNIVERSITY PRESS
The Edinburgh Building, Cambridge CB2 2RU, United Kingdom
40 West 20th Street, New York, NY10011-4211, USA
10 Stamford Road, Oakleigh, Melbourne 3166, Australia

Sue Atkinson    Sharon Harrison    Laurie Rousham

Illustrated by Sonia Canals

© Cambridge University Press 1998
First published 1998
Printed in Great Britain at the University Press, Cambridge

# I have played these games.

## Snake game

Front cover

I can count to

I am going to try to

1. _____

2. _____

I can _____

_____

date _____

## Snake game

Rules

- Put your counters on the tail of the snake.
- Take turns to spin the spinner.
- Move that number of spaces.
- The winner is the first to get to 15.

CAMBRIDGE
UNIVERSITY PRESS

ISBN 0-521-63423-7

9 780521 634236